The Ultimate Theory
of Everything

The Ultimate Theory of Everything

And Other Misguided Thought Experiments

Roy R Porter Jr.

Library of Congress Control Number: 2010910850
ISBN: Hardcover   978-1-4535-4361-0
                   Softcover   978-1-4535-4360-3
                   Ebook   978-1-4535-4362-7

This book was printed in the United States of America.

To order additional copies of this book, contact:
Xlibris Corporation
1-888-795-4274
www.Xlibris.com
Orders@Xlibris.com
84433

In Loving Memory of my Mother, Ruth L Krakowski.
Her support, kindness and long talks about matters including this
subject deserve recognition and thanks. Her two sisters and
my Father's Sister also know of her numerous,
extremely self sacrificing acts of helping
others she loved.

Preface

Ever since Albert Einstein first published his General and Special Theories of Relativity there have been many attempts to prove or disprove his theories. This book will not attempt to do either. Instead, it is a collection of ideas without mathematical mumbo jumbo whose sole intent is to engage you in your own thought experiments. There is no complex mathematics or long formulas to understand, only ideas to make you think and imagine.

Not only are Einstein's Theories used as a basis for the ideas here, but many other ideas from physics, chemistry, time and space are explored. This is not a book to educate you or explain the unexplainable, only a series of ideas that can only make sense if done as thought experiments. Once real Math and Physics are applied, many of these ideas may just fade away as improvable, unthinkable or just plain insanity. But this book is not about that either. It is about the process of thought and "thinking outside the book" as I like to put it.

This book is based entirely on my previously internet published essay titled:

"Universal Time Constant to Measure Gravimetric Time Distortions around Massive Objects and Calculate Actual Interstellar Distances" (Say that 3 times fast).

I started the below idea and paper over 15 years ago!

Universal Time Constant to Measure Gravimetric Time Distortions around Massive Objects and Calculate Actual Interstellar Distances.

(Say that 3 times real fast!)

(And why we don't need no gravitation)

(Poor grammar intentional)

Expanding universe, why won't it just behave?

Maybe because the universe is really standing still!

And why C causes shrinkage!

And to throw in a curve, is space straight?

(Space gone straight? Here is how to call it out and it complements the LIGO Detector!)

We don't need no education or maybe a change for the future?

Hypercube, 5 dimensions or 8? Unlimited Dimensions?

"They called me mad, and I called them mad, and damn them, they outvoted me. Quote from: Dr. Roy Porter "*Madness*: A Brief History", Oxford University Press

Time warps? Yes it does!

Not your normal F=MA!

It slows us down but makes us get pushed!

Dark matter.. Bah Humbug! We no need that too!

For that matter, I just about explained everything!

Well, almost everything!

Oh, OK, well maybe not, but fun to read!

FINE PRINT:

I have sent copies to many physicists worldwide including Stephen Hawkins! I personally handed earlier copies to Bill Nye (The science Guy) and Neil Tyson (NOVA) and scattered a few around the buffet table to the dismay of the other guests, at the Planetarysociety.org convention October 2007 at the NEW World Trade Center Building 7! (But they never even sent me a million dollars or a rubber chicken or anything like that.. Hmmm??) Reference: http://en.wikipedia.org/wiki/Gravitational_time_dilation

Although I have no degrees in quantum mechanics, physics or related mathematics, Science would benefit from a time "standard" in which to be able to mathematically calculate the velocity of objects near a massive object and distances between massive objects. Two possibilities as a time "constant" could be time rate at sea level or time progression far away from the center of gravimetric time changes caused by our own solar system. Voyager 1 is a candidate, if possible, being the only man-made object to be almost free of any gravimetric influences from our solar system and therefore the change in the rate of time flow should be almost zero.

Once this can be done, then we can more accurately use the equations of speed and velocity to accurately determine an object's relative velocity and actual interstellar distances when affected by a massive object.

First and foremost, gravity may not be a force; it may be the RESULT of this change in rate of time flow (see above link). We will arbitrarily say an object moving 60 ft/ sec at maybe 200,000 miles from sea level on earth will be our 1 sec constant. Therefore, since time travels slower at sea level (see above link), it would be the same as adding, maybe .001 second to a like object traveling parallel (negating air resistance etc.) at sea level. Therefore, let us begin with an object moving 60 ft / sec for our constant in space and the sea level object would be 60 ft / sec +.001 (time constant plus adjustment) or 59.940ft. in one second (constant) as time passes for our constant in space. Therefore, we see that the object at sea level traveled less distance or SLOWER USING AN ADJUSTED SECOND! (All numbers are not actual values since they would be even smaller, so larger numbers are only used for an example throughout this Essay and to prevent myself from becoming hopelessly confused also.)

This theory leaves other possibilities open such as:

- Gravity is a DIRECT result of time changes created by massive objects.
- Distances measured from Earth to other space objects are higher than if measured physically. (Light Pulse vs. Tape measure)
- The Shapiro Delay is proof?
- Space-time as a measurement!
- Law of Inertia explained!
- Objects of different masses fall at same speed, EXPLAINED!
- Gravity/Time dilation at single atom level?
- Light itself may be a form of time packets.
- General Theory versus Special Theory. Mass creates gravity, acceleration and slows time but C increases mass and slows time. Same variables! Just a co-incidence?
- The expanding universe may be closer than we think.
- The Grand Unification Theory may work if Gravity is removed since it is not a force!
- Possibility of Field Propulsion?
- Red Shift of Light from moving objects or gravitational influence?
- What really happens in a black hole?
- Explanation of the big bang theory!
- No need for Dark Matter!
- Magnetism is gravity?
- Why the universe seems to be expanding faster than it should be.
- Conservation of Motion? Oh no, if Newton ever knew!

First, let us use the analogy from earlier and add to it. Instead of an object in space moving parallel to an object on Earth, let' use a 10 mile long pole moving 60 ft / Sec. PARALELL but perpendicular to an object at sea level with the end farthest from Earth being at the 200,000 miles we defined earlier as the one sec constant. Then the farthest end would be moving at 60ft / sec (time constant). The end of the pole closer to earth would no longer be in the "constant" area of reference but be closer to the Earth's sea level object and therefore subject to a time adjustment of let's say +.00001 to offset the difference. Now the closer end would travel at 60 ft / 1.00001 or 59.9994 ft in one adjusted second, traveling <u>slower</u> than the far end. Would this effect cause what appears to be it going into orbit or arching toward Earth since the closer end slows down causing an arching effect. Does this sound like the behavior of smaller objects such as asteroids approaching near parallel to the Earth whose end farther away from the earth seems to travel a bit faster (or the near end travels slower which is more accurate) so it "arches" around the mass of the Earth.

You can use the same analogy if something is approaching directly toward the earth but you would need to take into account that even something directly approaching Earth at a 90 degree angle has to also be moving equally parallel to the Earths forward motion, which of course is rotating, orbiting the sun and orbiting the Milky Way with the sun, and moving away from everything in the expanding Galaxy.) It may be MOVING zero ft / sec parallel to the Earth but <u>not</u> zero ft /sec to the moon, sun, solar system or Galaxy or other space objects. Therefore, even though an object appears to be moving directly toward us, it is also experiencing the same forces of an object that moves only parallel. This "overlooked" movement has the same effect except the "arch" is not apparent to us and appears that the object is pulled toward the Earth but it is still following the arching effect as before relative to other objects in space. Therefore, this not so apparent arching also explains the effect of attraction or Gravity but with a different angular momentum.

Furthermore, since we already have discussed that the closer an object is to Earth, the less distance it will travel if compared to the time Constant. Well, light would also be affected in this way. Therefore, light would actually take longer and longer (nanoseconds more?) to reach sea level if arriving from space even though RELATIVELY it appears to always move at the CONSTANT speed of light. Light has to change its velocity too. If this were not the case then consider what would happen if light "ignored" the time "zones'". It would always go the same speed according to its own version of a standard second, regardless of time dilatation.

Therefore if we measured the distance to the moon with a laser, would it appear to be farther away taking into account the time differential near the two massive objects? If we tried to run a tape measure from here to the moon, would we find that it showed the moon was physically closer than we measured with a laser pulse? Once again, we would need a "constant" time reference to create the equations to accurately measure space distances using adjustments based on the "time constant".

Also, acceleration mimics gravity. We call it G-forces. Well, the acceleration equation uses a changing variable of distance traveled. Less distance traveled per time unit for deceleration and more distance travel per time unit for acceleration. Well, if instead we apply the changes of time as we approach a massive object and instead of changing distance traveled, we use changing time flow. We keep the distance traveled at the same value but instead use a changing time flow. This would also give you an acceleration value called gravity! Whoever said we couldn't change the value of T in $V = D/T$ and $A = \pm$ Delta V to give the SAME value of V and A?

Speed of light causes shrinkage? (But can add mass) Feb 06, 2010

More to consider: As an object approaches the speed of light, time slows down and its mass increases. Well, does time slow down BECAUSE its mass increases? Or because time slows down, its mass appears to increase? Mass causes time to slow down in our previous "thought" equations and also causes Acceleration in the direction of the Mass. So M=-Delta T and Mass equals +Delta A. So then using simple rules of Algebra, If M=-Delta T and M=+Delta A, then –Delta T= + Delta A!

But this is for Acceleration. Well since we are now at a constant Velocity near light speed, time has slowed and an object can't really gain mass according the laws of physics so what has happened? A Positive Change in Velocity (+Delta V) resulted in a Positive change in Mass (+Delta M) Therefore we have +Delta V = +Delta M.

Coincidentally, as a result in the +Delta V we also had a Negative change in time flow or –Delta T. So therefore +Delta V = -Delta T.

Once again, since +Delta V= +Delta M AND +Delta V= -Delta T then we can say +Delta V= -Delta T. Reverse it to read –Delta T=+Delta V. So which came first, the chicken or the egg?

Neither! What we have here is another variable not yet mentioned, the cumulative effect of acceleration and the –Delta T of the Mass.

Huh?

Ok, during the acceleration we have a positive change in the Velocity equation or simply +Delta V. We also have ANOTHER change in the velocity equation from the effects of the mass. M= -Delta T (however minute, depending on the size of the Mass). So we must add them together. Delta V1 of Mass being accelerated (say by a rocket) PLUS Delta V2 created by the mass (measured motionless) Therefore, this would explain the whole problem with accelerating a Mass and why it appears to gain weight, (or mass).

We have to add Delta V1 and Delta V2!

But wait, Delta V2 exists in ALL directions. So now we have another issue to really rain on our parade.

Ok, so let's just focus on 2 discreet directions like a car going down a highway to simplify this.

Frontend heading down the road and backend leaving the road behind or forward end, trailing end or Plus and Minus.

Direction one, or forward end of our Mass is in the direction of acceleration. It therefore still is going at +Delta V. But we also have to add its gravitation time dilation forward also. So we have to add, once again –Delta T (for Gravity) to +Delta V (for forward acceleration). Like shooting a gun forward, the velocities add. (For now, forget about using a light beam due to theoretical Max Speed C, it will be used for this at another time with surprising results, but not yet.)

But –Delta T goes in all directions, including backward or out the back of our car. But the Back end is also accelerating with the whole car too (otherwise our trunk and luggage would be left behind on the road!). But, it is leaving –Delta T behind as it goes, like shooting a gun out the back window, we now need to subtract.

So, we have to now subtract –Delta T which is – (-Delta T). This becomes +Delta T.

Oh, wait a second, what did someone famous say about an object shrinking in the direction of travel?

Killing 2 birds with one theory, um I mean stone, keeps happening the more insane I get here.

OK, back to the point.   Front end of Mass: (+Delta v) + (-Delta T).
                                Back end of Mass: (+Delta V) – (-Delta T).

Earlier we found that M=-Delta T but we added that to +Delta V! Wait, earlier I showed that –Delta T and + Delta V are interchangeable. So let's interchange, shall we? –Delta T + (-Delta T) {interchanged with +delta V) we get 2(-Delta T).

Also I showed that M=-Delta T but we have to balance the equation to read 2M=2(-Delta T). So after all this, we found that we double our Mass every time we double –Delta T. Guess what? Einstein was right! Acceleration causes Mass to increase. I showed that and I gave the rate of increase too! (Could it be that simple or am I delusional, I think the later is somewhat more plausible)

Oh, but wait, we forgot to mention the trunk!

Oh and you thought that we couldn't see what was behind us, huh?

Out of sight, out of time?

OK, now we have the trunk, I mean the trailing end doing its best to hang on to the rest of the car or Mass, accelerating also (+Delta V). But our mass is leaving behind some –Delta T (clean, green T, not exhaust smoke and CO2!!). So as with a bullet shot backward we now subtract what we are leaving behind, like a trail of bread crumbs.

+Delta V – (-Delta T).

So this is the same as adding Delta T! Now we can do this two ways to give the same effect.

First, we have the equation M=-Delta T so to balance it we need to say –M=+delta T. And we have a + Delta T, do we not? So therefore the Mass is shrinking in our trunk! Our trunk is getting smaller! But the front of our car (Mass) is growing twice as much (2M) as the trunk is shrinking!

So what did Einstein predict?

OK, it is growing in mass (2M) in the front, "check", shrinking in mass at the rear, Um, not a "check".

Shrinking in the direction of travel would be a "check". Hmm, not exactly what Einstein predicted but its how we look at it.

OK, let's look at it differently then. (Who says I can't fantasize here!)

Again, the trunk is: +Delta V – (- Delta T)

And earlier we stated +Delta V=-Delta T

So, let's do the interchange again.

-Delta T – (- delta T) or

-Delta T +Delta T

Which = 0 delta T!

Um, that's Zero Delta T, right?

So, there is NO Delta T from our mass?

And with no Delta T, we have no gravity in our trunk! Must be awkward or maybe even fun for any stow-away!

And from earlier we have 2M=2(-Delta T) under the hood! (What is that in CC's?)

Oh, boy, now our trunk seems to be unusually attracted to the strange increase in the size of the Engine under the Hood (2 times Mass or 2M, but that means 2 times gravity!). Also the trunk is now losing its own attractive power (0 Delta T=0 Delta V, gravity). For every unit of attractiveness the trunk looses the Engine size doubles (2M).

So, as a result, the trunk shrinks in the direction of the front of the car one unit for every time the Engine doubles its size (gravimetric attraction by MASS size).

Well, at least my insanity agrees with Einstein by sheer manipulation of the facts in that:

1. Mass increases with acceleration.

2. It also shrinks AWAY from the OPPOSITE direction of travel or better to say "shrinks TOWARD the direction of Travel".

As with all the thought experiments present in this essay, there are probably numerous ways to show this all to be rubbish but this is written for enjoyment and for you to be pleasantly bewildered, amused and confused. One example is the gain of mass here can be neutralized by the loss in gravity in the rear, but hey, I never claimed it foolproof.

A quick thought on single atoms. Since there is such a minute distance from one end an atom to another, would a lone atom be less affected by gravity/time? Thinking of the time dilation as circles in a pond, radiating outward, diminishing with distance, would there be enough difference in time dilation from one side of an electron or proton or atom to create enough of the effect described above. Would their size alone DECREASE the effects of gravity, one again causing many calculation and observations to be strange or inaccurate at the atomic level?

Let's bend the Law, Newton's! February 08, 2010

Oh, gee, and what of inertia? How can something in the middle of space away from any object seem to have weight just because we decide to give it a heave? Newton's Law of Inertia was of course right but did he explain why he was correct?

I just agreed with him too but now let's explain why it is correct!

How?

Well, after the thought experiment as to why an object seems to gain mass and that time and time and time seem to add to, um, more time and no time.

Just consider the same silly equations earlier about adding and subtracting –Delta T from anything in motion. It balances out once acceleration stops. No Delta V, nothing to add or subtract thus the stable, no motion state of 360 degree –delta T is all that is left. So, for all purposes, it is motionless in its own relative space..

Now just push on it a little (or pull if you like, have it your way, it's your thought experiment!)

Once you gently push on any object, the equations begin to reappear for both the Hood and Trunk of Car. The –Delta T now gets a boost from the + Delta V in A. Remember A = Delta V/ Delta T.

Oh boy, Now which is the Delta, V or T. Gosh,, But either way, that now has to be added as before to our ever present omnipotent, um, Omni directional –Delta T from the Mass. Oh and Guess what? The larger the Mass the Larger the – Delta T and the more it will resist a nudge.

As long as you understood written earlier here about adding the Deltas at the front of our car and then the rear, you can see the relationship to inertia with a little insane imagination like I use! No need to go over that again right?

Donuts and coffee anyone?

What if you take a super strong metallic donut shaped ring, rotate it to within a few percentage of light speed. Would time slow down inside the donuts center? If that occurred, could you then place an object near it and the time distortion would cause "acceleration" toward the center. If you then use that acceleration to push the donut away from the object and repeat it again, would you have field propulsion or a flying saucer? (Don't they always rotate at high speeds too? MMM that's an interesting coincidence . . .)

How? Use MAG_LEV technology! If the rotating ring is magnetized and a nearby same size non-rotating ring is turn on and off magnetically, we have the use of two weird forces working together to do something! Let's engineer that!

Oh, and does super cold substances tend to defy physics? Or maybe they only seem to. Maybe they are just obeying time, not disobeying physics! Slower, colder time!

Red Shift of Light from moving objects or gravitational influence?

Why does light "red shift" if it is emitted from an object moving away from the observer? Since no one has been able to invalidate C as the "ultimate" speed and speed of light in a vacuum, light must "Speed up to C" to "conserve" this constant. But there must be some kind of energy exchange for light to do this. It can't just magically "speed up" by itself. What if light "takes" from "time" to do this? Then we would have a source of the "energy" that light needs to achieve light speed. Well, as we all already know, the wavelength of light decreases or red shifts. But DOES THE WAVE LENGTH ACTUALLY CHANGE OR DOES THE RATE THAT TIME PASSES FOR THAT PACKET OF LIGHT DECREASE? Since Wavelength is a result of frequency and time interval, did light really reduce its frequency or "borrow" energy from that one second interval, making the second last longer? (This also leads to more questions like: does mass drain energy from time causing time to slow down?) Instead of changing the Frequency, did it instead change the LOCAL time unit? 60 Hertz is 60 times a second, so what is 60 times in 1.001 second? This would only be local to the packet of light that needed to borrow from "time" so it could achieve light speed.

But is light wavelength shifted around massive objects? If light just bends and appears on the far side at the same wavelength, did it shift down in wavelength then back up to normal as it left the influence of the time distortion? In this way, it would maintain C throughout the area of influence? But from what observation? Here again we have the problem of a "Time Constant" to calculate this. Did the light take longer to "emerge" from the area of influence then if it were a straight line? We can call it a "curve in space" to explain if it takes longer but was there really a curve? Was the light affected differently by external time shifts than if it "borrowed" from time as in the previous paragraph?

Finally, as stated earlier, if light is affected by these time "zones" or changes in the rate that time flows, then imagine the effect that a massive object such as a galaxy would have on how time flows. Would this not mean that there would be an even greater difference between using light or using a physical tape measure? This would mean that Galaxies may be much closer than we calculated!! Therefore maybe the universe is not expanding as fast and as far as we think . . . But once again, I do not have the mathematical skills to show or prove that. I will leave that to mathematicians and physicists who may take this seriously and do the calculations . . . It may be a TIME for changing our understanding of Gravity. Pun intended . . .

And now the Shapiro Delay!

Addition 21 January 2010

Don't have a tape measure to test this? Then let's use the Shapiro delay instead! How? First, we know that mercury circles the sun as does all other planets, right? Even if it is elliptical, I know scientists have already measured the EXACT orbit of Mercury, or at least calculated it. How am I sure? Because when Mercury was on the far side of the Sun they had to adjust for the EXTRA slowing of time as their radar beam passed by the Sun!

So what?, you may say, that was needed to make the distance calculations correct otherwise Mercury would appear to" jump" to a point farther out of it's orbit away from the sun. EXACTLY!

Huh? Exactly what?

If we didn't adjust for the time change due to the Sun's Gravity, Mercury would measure farther away than it really is!

That is the point! We adjusted for the Sun's Time distortion, now if we adjusted for Mercury's own time distortion, wouldn't that also mean that it is CLOSER than we thought. Oh, and then we also have our own Earth's time distortion to adjust for also. What's good for the Goose is good for the Gander, or better yet, what is good for the Sun's mass is good for ANY mass! Need for a time constant after all? Are all our distance measurements slightly wrong?

CAN WE MEASURE SPACETIME?

Space-time, what is it? Well, to begin with it is how distances should be measured! Maybe Einstein hinted at this but why call it a curvature? Could he possibly have known and only misled us, waiting for someone to unravel the simplicity of it, although complex?

To start, according to the Shapiro Delay, a planet, i.e. Mercury, will seem farther away if measured by a radar pulse that passed by the sun. What we do is adjust the calculations. Well, instead we can represent that distance by using an equation that uses the actual distance in space and the time light takes to travel there.

Well, why would we do that? First of all, Mercury may not be the idea body to use in this example, so use the SUN.

We know that the sun is 93 Million Miles away using calculations that DO NOT account for time dilation. So, if we want to accurately describe the distance, we would use the unit SPACE-TIME.

Huh? What?

We would have to calculate the actual space distance as measured by a tape measure using it's time dilation to reduce the "observed" SPACE distance as measured by radar. It will be somewhat closer, maybe 92.75 million miles, but this number alone could not be used to calculate travel time, ie, light. For this we would need another unit, TIME. This number would be the length of time light needed to travel the distance of 92.75 Million Miles. This ratio would have to equal out to our present value of 93 Million miles but the ratio of space/ time (space distance / time to travel that distance at C) gives the TRUE SPACE-TIME distance to the sun!

To make things more complicated, the SPACE/TIME distance ratio would change if we began approached the sun. The closer we get to the massive object, the less of a ratio. Of course any movement parallel to a large mass would produce an unchanging SPACETIME measurement to that object. (i.e. Earthbound (surface) measurements). So SPACETIME measurements on Earth for roads etc. would be totally unnecessary since the Space-Time values and old values will be identical when moving relatively parallel to the mass of the Earth.

Would these numbers then also mess with our understanding of the ultimate speed, C? I hope so! Maybe I can confuse myself enough to write something about that soon too!

Let's drag Newton into this insanity

EMAIL SENT: Feb 04, 2010

SUBJECT: Einstein, Newton and others proved right and then verbally assaulted.

Anyone Bored,

Attached is my latest updated insanity. It is easily understood by anyone in science and may prove fun to read. If you chose to read it, then I am not responsible for any side effects or long term need of psychiatric treatment or drug rehabilitation. (It is an ongoing essay that seems to just ask for more outrageous additions, and I have many more delusions to add).

Please feel free to send this to anyone that may find it interesting, amusing or that deserves the punishment!

And what about conservation of motion? Let's violate that a little too! How, well, let's look at it from Newton's point of view!

An object in motion will stay in motion unless acted upon by an outside force, or so I am told in Physics. Can't argue with textbooks right? Or can we? Yes, and we should! But just a little!

Let's think OUTSIDE THE BOOK! Um, or was it supposed be box?

Let's take the previous example where the end of an object nearer a massive object experiences a different rate of time flow than the far end. Well, if we did real life calculations including the small difference in the passage of time, it would show that the object slows down, or for our thought experiment, is lagging in velocity of the far end. Is the far end end pushing it or is it just following Newton's Law that it needs to maintain the same velocity?

Huh?, well, adjusting for slower time flow creates a problem here since gravity, an outside force, is acting on the end nearest the massive object but adjusted velocity equation shows it slowing down in the direction of the mass.

This violates Newton's Law, so to conserve that, would the object appear to accelerate so that the velocity equation, with the adjusted second, equals the same speed as the far end?

Once again we have another question that seems to be without an answer. How can an object's nearer end show a difference in velocity than it's far end (away from a gravity source) when time dilation is figured in to the equation? Conservation of motion states that the velocity should stay the same or at least increase due to the pull of "Gravity" not, slow down! So to follow this law, the object would have to have a higher relative velocity TOWARD the cause of the time dilation then its further side so both sides follow conservation of motion.

Addition Jan 13, 2010

OK, now what about that rumor that a feather and hammer fall at the same acceleration on the moon? (Men on the moon? Faked? Well if the myth-busters SYFY TV show can confirm it who can argue?)

Well, yes, small items (um, under say, 1000 tons) would accelerate at the same rate in a vacuum following this theory. Why? Because I said so? Of course not! Just perform the same experiment earlier using a solid lead pole and then a hollow shaft made of Styrofoam. Now do the thought experiment all over again. NOTHING IS DIFFERENT!

Since both have relatively little mass in comparison to the earth, their miniscule gravity field would not add or subtract to the experiment enough to be significantly measurable by our standards. This would change if you had a large mass like if you had used the moon instead.

Additional thought experiment on black holes and the Big Bang:

Now that much has been postulated about the already known distortion of time, what would then happen as a massive object becomes even more massive i.e. a black hole? To begin with, time, as we know it slows down near a massive object as demonstrated by the known theories listed in the above link. That part is not in dispute and has been tested. So now let's take it to an extreme level as we find in a black hole. Time will of course slow time even greater as we get closer and closer to the center of a black hole. Remember as stated earlier, physical distances appear relatively greater than they really are if measured by a physical tape measure. This greater distance will manifest itself inside a black holes event horizon, not as physical distances, but as the slowing of travel of that physical distance, creating, sort of speak, a time warp that has been already theorized as creating the condition that it "takes forever to reach the center of a black hole".

Revision/addition of 22 February 2008

Well, yes, that has already been theorized but has it been explained as to why this happens? This does not mean that you would not be torn apart by the time/tidal/gravity forces, only that the closer a particle/wave/energy gets to the center; the longer it takes to go one step closer. This has an effect of a relative stretch in space, creating more and more "relative" space due to time slowing but does not actually create more "physical" space as would be measured by a tape measure. It is almost as if matter/energy is being buffered in the past (compared to normal time/space) due to the immense time differences between times in the surrounding normal space compared to the time dilated/slowed area past the event horizon. A buffer is created, not by adding physical space but by slowing time relative to normal space. The effect is of bunching up matter in slowing time zones, compressing more and more mass the closer to the center.

Well, so now we have the makings of a big bang that is in fact NOT A BANG AT ALL!

The immense build up of matter in the time zones will eventually start to affect nearby matter as well, clumping together on their never-ending journey toward the center of the black hole. Masses will form, creating their OWN pockets of individual time dilations/slowing of time as well as attracting more mass. The larger the black hole gets, the slower time passes, the larger the "relative" space for matter to accumulate grows and the more clumps will form. Over time enough matter may accumulate in numerous clumps to be equivalent of the mass in a whole galaxy! We have the start of a universe forming, waiting to be born!

As this black hole grows, by the slowing of time and build up of matter in the "relative" additional time/space more and more matter from more distant areas away from the black hole will be attracted towards it from far reaches of space causing a cosmic "big collapse". Matter from light years away, will, over countless millennium; end up in the time-distorted area of the black hole. The time distortion expands further into "normal" space as mass accumulates also adding physical distance to the relative distance added by the time dilation. Slowly, all matter, even that not yet crossing the event horizon is caught in the time dilation of this behemoth. This in turn changes the whole time frame of the universe in which the behemoth exists. As this time distortion/dilation spreads outwards, all is encompassed and becomes now, a part of the influence of that black hole.

At some point, there no longer exists an area not slowed by the black hole which then leaves a paradox that needs to be broken! If there is no longer a "normal" time zone in space anywhere, then there can also be no "slower" (dilated) relative time zone either. This is an over simplification put into black (normal time) and white (dilated time) instead of shades of grey in between. This creates an imbalance in the whole fabric of space-time. How does that imbalance balance itself out?

In a simplified model, the time dilated space builds pressure as it buffers matter into the slowing time-space compared to ever shrinking normal flowing time away from the area. This difference in time can only exist if both sides, normal and dilated time exist. As matter is pulled out of "normal" time, and as normal space ceases to exist, something that once was there as reference to determine what is "dilated" time, slowly ceases to exist. This begins to create a time flow paradox in which a time "vacuum" or lack of "normal time" exists. Without this, the effect of the time dilation that creates the "gravity" of the black hole cannot be measured, therefore the time dilation itself begins dissolve as "normal" time is no longer there as a "reference" in which to create the time dilation. Gravity itself begins to redefine itself compared to the lack of "normal" time. As this occurs, the "relative" space created by severe time dilation slowly becomes the "new" normal time rate.

The "not so BIG BANG" which is more of a transition, begins to become the only time frame left in the universe. Since there is no longer a "barrior" between normal and "dilated time", what appears to be a "relative" expansion occurs filling in the void left behind by the now non-existent "normal space-time" The expansion is so thorough, that even the center of the black hole finds itself expanding due to the lack of a "comparative" normal space time to "balance" out the extreme from the normal. Without the "positive" pressure of existing normal space-time, the "negative" pressure of the black hole is free to expand, unheeded by the now non-existent balancing and restrictive, "Normal space-time".

Of course, this is all just a thought experiment. It may entirely science fiction and mathematically inaccurate but it is "thinking outside the books" that, I hope, may someday lead to a more accurate and scientifically provable theory as to how black holes operate and how our universe came to exist.

Another possible theory is what heat makes objects expand. I never did like the theory of atoms "bouncing" around and colliding off each other. It seemed that if that was the case, there would never be such a thing as a solid since the atoms or molecules are free to "bounce" or "vibrate" around so therefore should be able to bounce right away from each other.

Instead, what if heat energy is a change in the rate of time flow instead of a change in distance traveled (more distance in the same second for instance). Then each atom would be just a little "faster in time" as we saw earlier (theory only) objects accelerate toward "slower time" or away from " faster time" Therefore the atoms would be "faster in time" than any surrounding time which would make the atom want to go in all directions at once towards the slower time or "expand". The other nearby atoms in the solid would be at the same "time frame" it would equalize any expansion or attraction. So the outer edge atoms of solid would be slightly pulled outward toward colder space. The inner atoms would in turn want to follow the outer atoms but since the time is almost equalized between atoms, there is not enough difference to counteract the forces that held the solid together in the first place until the "time difference" reaches a point where the attraction toward "colder" Is almost equal to the force holding the solid together. Then we have a liquid or cohesion as in water. If the time change creates enough acceleration toward "colder" Slower time then you will get a gas. If the added change in time is extreme and rapid, you will get an explosion.

Here are some old "givens" arranged to show the same.

G=Gravity D=Distance T=Time A=Acceleration V=Velocity

Gravity is acceleration (according to Einstein) therefore let's set G=A

We all know that Velocity=Distance traveled in a period of time or V=D/T

Acceleration is a negative or positive change in Velocity or ±delta-V so A= ±delta (change in) V (delta-vee)

OK we know basic math and that if a=b and b=c then a=c. Then we can say that:

G=A

and

A= ±delta-V

Therefore replace A with ±delta V

G= ±delta V

TRUE?

And V=D/T

So replace V with D/T:

G = ±delta D/T

OK, sure what is the point? Well, according to the General Theory of relativity, time slows as you approach a massive object due to gravity. Therefore, we need a -delta T, and then let's use that instead changing the equation to – delta D/T Since the + or - means CHOOSE one. So we did!

We now have G= – delta D/ T

And WE KNOW THAT gravity causes time to slow (negative) so we fix this to read

G= D/– delta T

Once again it all leads to Gravity being caused by the change in time. Too simple? Maybe that is why we keep missing the point! So simple that it eludes us!

WE DON'T NEED NO DARK MATTER!

Revisions/Additions November 2008

Well, as we all already know one thing without a doubt: Where there is Mass there is Gravity *and* time distortion. Sure, what does that have to do with dark matter? Well, for this we don't even need Gravity to be caused by time slowing. But we do need to TAKE IN TO ACCOUNT time distortions.

So, we have all this mass in the Universe and Scientists say it is NOT enough to account for all the Gravity, or also to say, there seems to be some mass missing so that would mean that with the missing mass there would be missing gravity. Or maybe there seems to be too much gravity and not enough mass to support it.

Well it is the latter. Scientists are looking for enough matter to equal the gravity we measure. Well, THEY WILL NEVER FIND IT! Why? Because the matter IS NOT MISSING! So what's the matter with the Matter already there? What they forgot to figure in is that the gravity CAUSES time dilation. WHAT? SO? Well, even if time dilation is just a byproduct of gravity, then it still explains the amount of missing "gravity like force"

Let us now go back to acceleration just to get your mind in the right space. Gravity and Acceleration are indistinguishable. Well, what is Acceleration? It is a change in velocity. And what is velocity? Velocity equals Distance / Time. (V=D/T) So if you change (T) time (time dilation from all the mass in the universe) you get a change in velocity. And what also seems to be exactly the same? Gravity!! So, once again so you understand how simple it all really is: Gravity equals Acceleration equals change in velocity equals Distance/Time Dilation, A=B=C=D is the same as A=D? ARGUE WITH THAT if you dare . . . . Or don't argue, no pressure . . .

Now take the extra "gravity" produced by this and add it to the all the gravity in the Universe and do you then have the missing numbers instead of the need for dark matter? Well, someone do the equations and find out please, I am dying to know ☺

The Universe is really standing still!

Added May 31, 2009 My Birthday!

We all know that Galaxies are massive! But since we already know that mass slows time, then how much effect would an entire galaxy have compared to our tiny reference here on Earth? Furthermore, what effect would that time "zone" have on our measurements? Since time slows, would that "stretch" our observations in the same relativistic way and the only way to accurately measure the mass would be to adjust for the time difference or be in the same time zone? In other words, are we normal time trying to make sense of readings taken during daylight savings time without knowing there was a difference? Our calculations would be off! No one told us that the galaxy we are measuring is on daylight Savings time! Would we then try to invent all kinds of theories to explain it such as dark matter, 11 dimensions, string theories, gravity waves, and gravity photons? Are we trying to figure the 3rd shift payroll on the Sunday of Daylight savings time without knowing it was daylight savings time?

What is magnetism? What happens when an object travels near the speed of light? What happens as electrons orbit at the speed of light around atoms? Well since most electrons are in pairs spinning in opposite directions, would that create a net external speed of zero if one direction caused a "left" amount of Special Relativity and the other in the opposite direction caused a "right" amount? Would the "left" and "right" cancel each other while also creating a combined "local" speed of 2 time C! Would that cause all kinds of weird readings like being in 2 places at once, appearing to be a cloud or then a wave? And what about the time slowing and shrinking in direction of travel? Poor electrons must really have a heck of an existence!

What happens if there is not a pair as in magnetic items? Well then there is small local "f left" time dilation and mass increase. This dilation is directional as in the theory of the "spinning donut" with the electron being the donut. Line up the donut holes and you have a magnet, with two poles, North and South (or Left and right?) Since it is so minute, it has almost no effect on anything else except, except . . . .

Except maybe . . .

Maybe!!!

Maybe except for another electron that is also without a canceling partner and is a spinning lonely donut with two poles. Move the two close enough together and the combined time difference creates an effect similar to gravity as explained above. In addition, physics tends to create a neutral stable state. Magnetism is also the electrons attempting to get together to create a net of zero?

But you may ask. "Well what of lightning and free electrons"? Simply, they cannot form the necessary donut ring since there is not am atom to circle. Now if I could only explain why they like to make donuts around atoms (Even if the donut changes orbit, it is still a donut. Any other orbit would not be an orbit would it? Imagine a satellite that does NOT form an almost perfect circle (its orbit can change slightly with the maximum deviation being dependent on the mass, speed etc) It wouldn't be a satellite but a space ship going somewhere else if it wasn't making a "donut" or close to a "donut".

Now let's make light into something really weird too!

And as we all know when light is absorbed by an atom, it excites it or makes it hotter etc. Hot is described as atoms "moving" faster or vibrating which is known as higher temperature. Well, now seeing that the speed equation can be adjusted on the "time" side as well, are the molecules traveling farther in a second? That is to say would an atom that absorbed light go from "vibrating" 60 Pico meters /2 nanoseconds (arbitrary numbers) to 65picometers in 2 nanoseconds (more distant traveled in the same measured time relatively)? Could it in fact be 60 Pico meters in 1 nanosecond instead, as the light packet releases its stored time energy, making time itself speed up in a subatomic area? (Remember, locally it would still read 2nSec but an outside observer sees that time pass in 1 n Sec. (Confused? LOL, So am I sometimes!)

Do we ever see observations like this in decay rates? (Please for God's sake, you tell me, because sometimes I am the confused one ☺.) Could this also explain the experiments that show light as a particle and a wave depending on the measurement and circumstances? What if it were packets of time? (Now that is an idea hard to bring to light). But all this is above my ability to test mathematically. I would welcome any comments from scientists and mathematicians that could do the equations and calculations.

If light is affected by these time "zones" or changes in the rate that time flows, then imagine the effect that a massive object such as a galaxy would have on how time flows. Would this not mean that there would be an even greater difference between using light or using a physical tape measure? This would mean that Galaxies may be much closer than we calculated!! Therefore, maybe the universe is not expanding as fast and as far as we think . . . But once again, I do not have the mathematical skills to show or prove that. I will leave that to mathematicians and physicists who may take this seriously and do the calculations . . . It may be a TIME for changing our understanding of Gravity. Pun intended . . . A

Addition 24 April, 2009

Why doesn't the universal expansion add up? There should be enough gravity to create a "big crunch" but that doesn't convince the universe to follow those rules. What's up with that? Once again, did they take into consideration that time runs slower closer to a galaxy? Add it all up and what do you get? Well, a bunch of stars that just won't behave! Well, can't blame the stars, they just can't keep up, literally! As the seconds tick away in empty space, a few milliseconds are lost. As the years tick away in empty space between galaxies, (like, aren't we in one of those?) seconds are lost for the galaxies. So, without taking the cumulative loss of time for each galaxy, are we making a cosmic blunder? Once again, if a tape measure were available, we could accurately measure the expansion rate, but we don't so we use light that crosses the empty "normal time zone" space and then enters our slower space. Well, if we compare measurements from decades ago until today there will be a few milliseconds (?) added to the cumulative slowing of time.

So once again using the earth-moon example, what would the difference be if we made a measurement with a light pulse from here to the moon, recorded that. Then, we shot a second pulse and to simulate a few thousand years of cumulative slowing of time we purposely add a few milliseconds to the time the light pulse's travel time. Well, we now think the moon is even farther away! And of course, I don't have the tape measure or the laser to prove this so I leave that up to the ones with Government funding. But please invite me and let me in on a little of the funding and fame, OK?

Now let's combine what was said about black holes with what we were told about the universe, that it is expanding. Well, is it really? Or does it just APPEAR to be expanding. Take a black hole, buffering matter into the past due to the intense slowing of time as we approach it, now take the gravity of a galaxy and would it not be MORE? Well, what does that prove? Not finished yet... Now remember time dilation and accumulated slowing of time. Would that not also create a similar buffer around ANY object with mass. But since the distortion allows light, radio waves to escape are we seeing them coming from the past, not from farther away? In other words, it is taking more and more time for anything to leave the vicinity of a gravity source, therefore distorting OUR perception of when it really left, giving us the WRONG perception that it is moving farther away but it is actually in the same place but moving farther back in time compared to normal space. So it would then take LONGER for ANYTHING to leave that area due to accumulated time dilation but we misinterpret that as farther away in space, not farther away in time! So how much is the galaxy ACTUALLY expanding and how much is an illusion from accumulated time dilation around massive objects? If anyone can answer this, please feel free to maybe let me know!

Newton's Inertia! Why can't it just stop?

And how to represent time dilation, Use the Force!

EMAIL SENT: February 13, 2010

SUBJECT: A disturbance in the force, to explain Newton's Law of inertia and Time Dilation.

Dear Yoda,

May the force be with you and anything else with mass! I am referring to Time Dilation Lines of Force seeping from all matter and how it affects everything. Latest update added at the end of the attached essay, re-formatted and more typos corrected.

Enjoy it or destroy it. No hard feelings.

Roy

Earlier I explained that an object at rest stays at rest due to adding of time dilation in one direction and subtracting in the other direction. Now to explain why an object doesn't want to stop!

Once again we have to take in to account that all mass emits time dilation in all directions. Now, we have to take "one giant leap" for mass-kind. (Oh no, no more puns or butchering of famous sayings, please!)

First, all mass is in motion, right? But compared to what? Here we have to take a giant leap and isolate a mass in empty space far from any other mass. Now we have to also say it is moving, but relative to what? Or can we say it is standing still, once again, relative to what? Now hold that thought, um or that un-thought or whatever it is.

Now, think about what is known in science as lines of magnetic force. But wait, I never said it was magnetized, only to think about lines of magnetic force.

And we know that certain things happen when something changes lines of magnetic force, so here; let us think of Time Dilation as Dilation Lines of force.

Now, as discussed earlier, when an object is pushed, - Delta T is added to the front and subtracted from the rear. Instead let us think that Dilation Lines of force are bunched together in the front and are separated or stretched apart to the rear. This will make the whole thought experiment easier to understand.

Now, the object is on its merry way after we gave it a push. Now no acceleration is added so it is just drifting away. Why won't it stop? This is where the Lines of Dilation Force come into play again. With no further acceleration and NOTHING nearby to interact with the lines of Dilation Force, they once again begin radiating evenly in all directions. Since space is flat (I don't believe it can be curved.) and there is nothing to disturb the lines of Dilation Force, the object and the lines of Dilation force now are at a stable state that for all intent and purposes, is that of a motionless object.

But as we stated earlier, without a relative reference, we do not know if it is motionless or racing by. Once we measure it, WE become the reference. And to disturb it in ANY way will now cause a disturbance in the, um, force, Luke. Dilation Lines of Force to be exact. This will cause a change in velocity or lack of velocity depending on YOUR point of reference.

Once again, this agrees with Newton's Law that "Once in motion, will stay in motion unless acted upon by an outside force."

But to explain why this Law WORKS; once in steady motion, a mass and its Lines of Dilation Force will return to the same as at rest, unless something disturbs its Lines of Dilation Force, which will then cause a change in motion. (By adding −Delta T to the front and subtracting −Delta T from the rear)

Conversion to a Time Constant and its results

EMAIL SENT: February 14, 0020 hrs 2010

SUBJECT: Delete if you hate the puns, bad jokes and the idea of a Time Constant.

Added a few paragraphs to begin explaining the results of using a time constant.

By now a few of you may have noticed one very important inconsistency that will need to be corrected as I add more to this whole mess. It arises from my feeble attempts at not confusing myself, and guess what, I got confused.

First of all, if you wanted to describe a "slowing" of velocity you would indicate – Delta V. So if you wanted to describe the slowing of time, would it really be – Delta T. The answer is, surprisingly, no.

The reason can be directly derived from the title of this essay-minibook, specifically the term "Time Constant". My whole thought of defining a Time Constant was to more accurately and easily plunge into the problem of Time Dilation.

So, the way to accurately and easily define "Slowing of Time" compared to another reference (i.e. Between the Earth and a Geo satellite) was to ADD to a base of a one second constant, therefore indicating that 1 second is now longer than before.

In other words, again, if our 1 second Constant at 200,000 miles away from the Earth (with the Moon on the far side of the Earth), then to indicate a slower, longer second, we would ADD to it.

So, the more accurate description of the previous examples would be that all instances of negative (-) and Positive (+) Delta T actually need to be reverse.

Since this is only an essay, I do not wish to go back and correct it every time but to instead use it as an example right now of why a time constant may be necessary to better understand Time Dilation, its effects and the comparisons of HOW we describe it mathematically.

The next few additions will be to re-visit what was stated before using the idea of a time constant and the correct values of Delta T. This is still a work in progress. I still have many more ideas to add that haven't specifically been addressed yet and other ideas here that I only scratched the surface on. Expect dozens more pages and ideas!

And to throw in a curve, is space straight?

EMAIL SENT: February 24, 2010 sometime after midnight.

SUBJECT: Someone threw us a curved, space and how to test it!

Dear victims,

I added an idea (at the end of the paper) on how to test if space is curved, with NASA cooperation... um, not likely, but hey, ideas start small and then disappear... um I mean develop into something bigger.

Me!

OK, so I lied. Well, not on purpose.

This addition is not about the correct values of Delta T, but a suggestion about a faster, cheaper, quicker NASA way, to test if space is curved.

Let's engineer this test using engineering accomplishments from the near past. What we need is:

1. A small solar sail like the one built by the planetary society but lost before it was deployed.

2. A very long (As long as possible!) thin strong tether attached to the solar sail, measured precisely to the millimeter or better, like the one NASA used in past experiments, maybe of carbon nano tubes, fibers etc.

3. A tiny probe with a very precise clock, small cheap on-board calculator, solar panel, solar sail, laser and relatively weak radio transmitter / receiver and a radio link for data exchange with Earth.

4. Another tiny craft with another very precise clock, solar sail, laser sensor and radio receiver transmitter attached to the far end of the space-time line.

5. (Depending on budget, as if we already have one!) To the trailing probe we can add: another data link to Earth and a second return laser with a laser receiver/sensor on the main probe.

6. (Depending on Budget and Insanity) Additionally, another solar sail spacecraft at a 90 degree angle of the Main (and trail) tethered to the main probe. By simply angling the solar sails slightly away from each other to cancel out lateral movement and pull on the main probe, the second tether can be kept taught and perpendicular to the test tether. This way a 90 degree "control" can be established. This tether is used for perpendicular measurements.

7. Biggest necessity: The insanity to try this.

Easy enough? Well I could never afford it. But here is how you would use it:

Launch the whole package, wrapped tight, toward the sun (solar sails works best near a sun!). When it reaches its closest safest distance to the sun (radiation etc.), use gravity from a planet or the sun to change its trajectory to head AWAY from the sun.

Deploy the main probe w/solar sail, laser, and radio transmitter with the tether attached.

Due to the slow acceleration of the solar sail, the tether would be gently unwound or pulled from its storage. The "trailing" receiver would at some point, unfurl its solar sail to "almost" match the pull of the first solar sail to avoid a sudden "jerk" when the tether is completely unfurled otherwise the tether would need to be extremely strong.

Once the tether is completely unwound the trailing probe would close its solar sail, only to be dragged by the tether, therefore ensuring it is taught and straight.

Once this has been assured, the two probes, hopefully dozens or even hundreds of kilometers apart, now have a precise tape measure between them. Now we start sending radio messages and laser pulses between the two and, using the KNOWN length of the tether, we compare the travel time to the ACTUAL distance as measured by a tape measure! But . . .

Since the whole mess is being dragged away from the sun, we can see the REAL effects that proximity vs. distance has on space, time and whatever more we can think of the add to the experiment.

After time we can see if traveling away from the Sun's effect on space-time changes the readings and calculations of um, space-time. If, for example, the distance between the two crafts changes as measured by a laser although it is SET precisely by a tether, then we know something needs to be explained and we now have the data to calculate an explanation!

If I had the money I would love to try this myself. But I don't as an unemployed student using the New GI Bill at the Manhattan Campus of NYIT

I would have volunteered to go to go to war more than I already have been if I had known the New GI Bill was coming! And after this semester, I may go back to the front lines as a civilian technician or associate engineer. I already partially listed my Army accomplishments on www.togetherweserved.com.

I though outside the book continuously and improvised or went above and beyond what was asked of me or expected of me. This caused others in competition to try to discredit me, minimizing my accomplishments or causing my accomplishments to be overlooked entirely. Win at any cost? Some out there still don't play by society's or moral rules and then tell you it's fair. Sad that the world is still like this at times. But our world has always been like this in one form or another throughout the history of Human-kind.

Any comments?

Feel free to email me or send insults since that may help too.

We don't need no education or maybe a change for the future?

EMAIL SENT: February Fury 26, 2010 11:00 P.M.

SUBJECT: Importance of adding a time constant to education..

Recipients,

Here we go again! I added a few paragraphs stressing the implications of adding the theory of time dilation and a time constant to advanced physics. My essay is an example of the weird and possibly (or implausible) important ideas that would result. Time dilation is a fact that I believe can't be ignored!

What will be the final lesson in all this? Education! We have been taught to only use the velocity equation (Distance/Time unit) with only the distance as a variable and a second, minute or hour as a universal constant. Until we start teaching that the denominator is a variable also, we will remain stuck! In other words, we already know that time units are not un-variable. One second at sea level is different than one second at Geo-stationary orbit distance.

If we want our future generation to be able to understand and subsequently develop a better understanding and implications of time dilation then we need to add that to advanced college physics (4th year plus?). In this essay I barely began to scratch the surface of the possibilities and theories that will come to light as a result of an understanding, however limited at my skills, of the implications of considering a time constant.

Establishing a REAL time constant would simplify our future endeavors at understanding and then teaching the future generations. In other words, once we have a time measurement standard, possibly even calculated at 0% gravitational influence, we can then use the velocity equation with the denominator as an EXACT number value of that time standard, which would more accurately represent time in areas that are not at 0% Gravimetric Influence (Time Dilation).

I truly believe that this step has to be taken so that our future will come closer to the theory of everything and advancement in all areas of Physics that deal with Gravitational Influences. Maybe it might eventually lead to the ultimate sci-fi fantasy of FTL.

I have an excuse!

EMAIL SENT: March 01, 2010

SUBJECT: I found an excuse!

All,

What we have here is a failure to speculate! How can we make time dilation calculations using our own, time dilated second? Our time is already somewhat outdated! My excuse for now!

Roy

So why doesn't the math work out for gravity and time dilation? Well, I can make up an excuse hard to disprove or prove until someone does what I ask first! (Calculate a Time Constant or "Zero G second") LOL, Is this called manipulation? But that is not the intent here.

All the calculations being done today on NEO's, planets, and satellites or such are done INSIDE already existing Dilation Lines of Force. The Sun's massive gravitational influence, of course, reaches everything in the solar system. (or else it wouldn't be a solar system, right?). So tell you something you don't know . . .

If, as I suspect, space is not curved, warped or twisted then there only exists a positive t effect throughout our solar system. (One second plus .01 second is longer, so it is adding or a positive number to represent time dilation) If you do equations that describe the slight change in effect that time dilation has, you are using a base of one second that is already dilated! In other words, saying the math is not right to give the right numbers is because it is still based on a NON-standard second. Every time I try to calculate the effect of time dilation I try to convert to a positive t value such that time dilation would equal 1 second plus (+ t) to represent a "longer" second. But that second is already longer than at Zero G!

The problem is that the 1 second reference I use is already been dilated by the sun, the Earth, the moon, space junk etc. So, until someone much smarter than me can calculate the actual rate of time passage at Zero Gravity, the equations will not work. (My excuse!)

So how in the world, or out of this world, can we possibly calculate a "base" second to um, base all dilation equations on?

How? Use acceleration! Simply, we can derive a very close approximation to a "base" or "zero G" second by inter-relating the acceleration needed to stay a set distance from the sun, add to that similar tests based on other influences (Jupiter, nearby planets, moons) and we can eventually calculate a close to accurate Zero G Second. Then let's see if we can play with the numbers based on that!

Recess!

Let's go play!

And of the LIGO Detector . . .

EMAIL SENT: March 06, 2010 2220 Hours

SUBJECT:

Anywho,

After seeing a video about the LIGO Detector (again?) I realize that my space based design follows almost the same design. Is it a Coincidence? Probably not, since I always watch shows about this type of thing. The point is that, if space is straight, the LIGO detector wouldn't measure a change in the distance, but a change in only the flow rate of time, hence appearing that the distance has changed. The LIGO detector is on the same path that I am, even if I stray a little.

CUBE SQUARED ?

EMAIL SENT: 14 Mar 2010 2020 Hours

SUBJECT: Dimensions found in unlikely places!

All,

How many dimensions in a cube? Hypercube? Space? Around a sun? More stretching of the truth and space..

Education is having a bad effect on me! Studying Digital Logic and the use of Cubes and Hyper cubes in Logic Minimization, I am straying further down the road to insanity!

How many dimensions does a cube have? 3 right? Wrong! Four, counting time! Then a Hypercube should have five? But as my Father put it today, it should have 8, 4 for the outer cube and 4 for the inner. It just boggles the grey matter.

So, if this is the case, then it adds to another theory that there are theoretically an unlimited amount of dimensions and parallel worlds. All we have to do is consider time dilation, or Dilation line of force as an additional dimension. Since there are many lines of force with Dilation would that then give anyone the ability to say there are multiple dimensions from Dilation? Overlap those with lines of force from a nearby massive object such as sun and then you have multiple parallel dimensions all based on lines of Dilation Force from another object.

So, to grasp at straws or stars... Did the physicists discover this and assume it to be something entirely different? Is their mathematical discovery of more than 4 dimensions just their stumbling onto this as I have tripped over it? Well, just more strange, inappropriate and probably implausible ideas to come from time dilation.

The more I learn at college the more I will come up with. I still have other past ideas that I haven't even discussed yet! College IS dangerous (to me at least)!

EMAIL: 17 Mar 2010 23200

SUBJECT: It started as a joke but beware the uneducated!

All,

    As far as I know I am NOT related to Dr. Roy Porter and my "Madness" has nothing to do with his books on that subject. (ha ha ha, very funny!) My ancestry does originate in England on my Father's side and his Father was an Engineer at G.E. in Connecticut. My mother's side comes from Poland and Immigrated to Kingston NY through Ellis Island.

    A joke did start many years ago because of Oxford's Roy Porter and the uneducated took it too literally, calling me "MAD" when the more educated had laughed and called me the "Mad Doctor". Well, which will you be? I almost went insane until I realized the punch line. Just search the internet, if you dare..

NOT Dr. Roy Porter

March 18, 2010 1130 Hours E.S.T., Earth Time (Time constant +???)

Email sent: 18 Mar 2010 1230

SUBJECT: A dip in the pool to explain a warp dive.

All,

Can matter instantaneously accelerate or "WARP" into motion. It happens all the time, time and again and time is the cause.

Pool anyone? The weather seems almost nice enough. And what of the conservation of "momentum" from one object to another like in the game of pool? Why do equal masses so nicely exchange velocities when they collide?

Once again, we need to again imagine Dilation lines of force. Although their emanation or "propagation" from a mass are so fast that they can almost be considered "instantaneous", we still need to assign a value, so for now let's just call the speed of propagation of the Dilation lines of force "velocity D".

OK, shoot the cue ball, with no spin, across the pool (table) to hit the eight ball straight on. Once the ball is in constant motion with no further acceleration, its Dilation lines of force equalize in almost an instant so that the emanations appear that of a ball at rest. Once again, assuming space is flat and has NO influence whatsoever on mass; the mass once again has an even "dispersal" of lines of Dilation. If it were not for another mass to compare to, we could not tell, measure or even know if the cue ball is moving. Space therefore seems flat and dimensionless in a case of a completely isolated moving (or stationary?) object.

Crack! (Not the evil drug), The cue ball hits the eight ball, which now almost instantaneously takes on the momentum of the Cue Ball which now has come to a complete stop? Why?

First, as equal small masses approach at constant velocity, dilation lines of force from each propagate outwards at speed "D" plus speed (or lack of) of each mass and cross each other, evenly adding to create a very tiny, negligible attraction due to time Dilation (Gravity) but the point here is that the lines are spaced evenly and a constant, even pattern is formed.

Upon collision, the mass in the cue ball undergoes an extreme deceleration. The Dilation lines of force are for a moment, out of equilibrium. A small gap develops in the forward lines of force. O)) "Gap"))))). This gap has to be accounted for by a loss of +Delta Time, therefore, for an instant there is close restoral to "normal time flow" in this gap. (Hmm, can this be used later to calculate an almost exact "base" Time Constant, but let's not go there now)

Remember, the lines of force are +Delta Time or Slowing of time. Once again using a reference time (Time Standard) PLUS the dilated amount or +Delta T gives you the equivalent of a "slower" time frame. Yes, it's confusing but if you consider the ratio Meters/Second. If you want to decrease the ratio (slow down) you would lower Meters OR, (mathematically correct also), ADD to the Seconds. Either (but not both simultaneously) results in a DECREASE in overall m/s.

So now that gap from the Cue ball has removed a line of force that would arrive at and combine with the Eight balls line of force toward the cue ball. The eight ball now has +Delta T lines of force that instead of combining with the approaching Cue balls + Delta T, are instead being submitted to a "GAP" or lack of +Delta T. This gap would have close to the value of zero Delta T. So, this gap continues at the Cue balls Velocity plus D across the Eight Ball, creating an extreme in balance until it reaches the other side of the eight ball where it again messes with the eight ball's +Delta T lines of force.

The overall effect is easier to imagine if we cancel out the velocity "D" of each ball. Instead we use the "gap" at the speed of the Cue ball. The "wave" of "un-dilated time" or "Gap in +Delta T" passes through the eight ball creating a type of travelling wave that seems to instantly change the velocity of the ball by "CHANGING it's time frame" but in a wave from one side to the other. In other words, it was "TIME WARPED" (from one side to another, sequentially) from its resting state to its final velocity after the collision due to the "non- dilated wave" passing over it.

So why does the other ball stop? We have to now reverse the whole process using the eight balls seemingly SUDDEN jump to its new velocity. This causes a similar but reverse effect on the Cue ball with dilation lines of force, causing the cue ball to "warp" to it's now seemingly motionless velocity.

So here we have it, warp dive! Well, more like a Giant "leap" into the pool.

I had an excuse! But it wasn't good enough!

The excuse was because the "acceleration" caused by time dilation seemed too insignificant to account for gravity. Well, thinking in normal physics I overlooked one significant difference. To accelerate anything, we apply the force to one end and it "radiates" up through mass to accelerate the farthest end from the force. This IS NOT THE CASE!

In other words, the time dilation cause of acceleration near a massive object is applying force to ALL parts of the object at once! So, using the example of the long pole at the beginning of this essay, it is not only the end that is being accelerated but rather the SUM of ALL the particles in the whole length of the pole. EVERY atom is experiencing the change in velocity from its forward side facing the gravitational mass according to the velocity equation, slightly changing the length of a second on the side facing the gravitational influence.

Add to that that there is the force that keeps mass together. By ALL the atoms MOVING spontaneously in the same direction due time dilation, the SUM would be an incredible magnitude higher than anything yet explained.

An example would be hot air and Hydrogen. Why would a hydrogen balloon float? Because the other heavier atoms of Helium down through the periodic table would experience a higher SUM of acceleration due to time dilation! Helium has TWO protons and TWO Neutrons, so would that be 2X Hydrogen's proton gravity plus TWO times Hydrogen's Neutron Gravity? (Forget about the electrons for now!)

And what of hot air? Already I pointed out that it may be that heat is actually caused by speeding up time. Well, then this is the EXACT opposite of time slowing down, therefore creating anti-gravity through heat alone. Increasing the flow rate of time creates heat, slowing it down creates gravity. So, by increasing the flow rate of time due to heat, the slowing of time caused by gravity would be cancelled out, causing hotter air to float.

Oh, and I have been researching some ingenious attempts to use similar methods to achieve field propulsion. Even the circulation plasmas still do not channel gravitons; they only change the rate of time flow. Similar experiments still do not take the rate of time flow into consideration.

Well, how do SUM all the effects of time dilation on trillions of billions of atoms, protons and neutrons? Well, that is what supercomputers are for! I would love to see the result!

Email sent: May 27, 2010 11:06 P.M.

Subject: No excuse, just a challenge!

Anywho,

What about the SUM of time dilation on every proton, neutron and electron? F=MA but ALL AT ONCE from time dilation? Changing the flow rate of time, albeit at a miniscule rate, across trillions of billions of atoms at once would add up to something, right? Anyone got a supercomputer to check this out? I left mine on the bus . . . .

Email sent: June 24/26, 2010 1:41 A.M./3:50 P.M.

Subject: Gravity, the simple solution is to consider the opposite!

Dear Anywho,

So, now let's try something completely opposite! Gravity decelerates us and causes other things to push us toward it! No that seems totally backwards! I added the full explanation to the end of my ongoing essay that Gravity is caused by Time Dilation. Well, read it and you decide!

As always, if someone published this earlier, then they deserve the credit if it has any truth.

Well, after all this, while half asleep, I seemed to hear an explanation that seemed to make sense of it all except it is the opposite way of looking at it!

But it still falls into line that Gravity is caused by Time Dilation! Not again! Not another shot in the dark. But it makes sense if it is done as thought experiment. Yes, it is still Time Dilation but this time it seems too exactly easy to dismiss.

Gravity does not accelerate us, it decelerates us? OH NO NOT AGAIN! Yes, as one end of us is nearer another mass, that end is experiencing a slower second, and if we take that into account in the equation of velocity, that end IS slowing down. HAH so that proves that it is NOT Time Dilation, right! WRONG! If one part of us slows down in one direction, the rest of us PUSH that part in THAT direction. See, we are slowed down then given a push all due to Time Dilation. Believe it or not!

Email sent: July 15/26, 2010 1:41 A.M./3:50 P.M.

Subject: Combining Einstein's General and Special Theories with Magnetism!

Dear victims,

   GREEN PROPULSION?

   If we combine Gravity, Time Dilation, Speed, magnetism and modern electronics, what could we create (other than something that might explode, warp time or swallow the Earth in a black hole).

   I added some additional experiments and tests in the "Mechanical Description" toward the end of the attached paper that may lead to some fundamental discoveries if not another possibility of field propulsion.

Roy Porter Jr

   Additional consideration in all this insanity has also created more insanity! The device that I described in the "Mechanical Description" also has other possibility for testing more ideas or just verifying old ideas. Using a magnet to push away another magnet is everyday physics although it cannot be used except in motors that use more energy than they produce in movement. Well, at higher speeds could that "push" be moved to another area, resulting in levitation or field propulsion?

   Once again, this idea could only be tested in the high speed platform I propose. Could applying a repelling force from the bottom of the containing platform to one small area of the rotating platform be transmitted a slight distance by the high speed rotating platform? If this "energy" or "motion" is offset just a little, could it then be used to repel another magnetic surface in the top of the containing platform, therefore creating lift? Field propulsion?

   Once again, adding time dilation from the high speed mass may increase the overall lift creating a feasible addition of forces? Once again, if one end of a mass experiences more time dilation than the far side, would conservation of motion cause movement to equalize this? Time dilation changes the value of one second, which changes the value of the denominator in Meters/Seconds which changes velocity! So what happens if one end of the mass "slows" down by this change? Does the other end "push" it back to the correct velocity value of that "time zone" therefore maintaining conservation of motion? Add this all up like "lines of magnetic force" for the whole mass, line for line of force, slowly decreasing further away from gravity or any time dilation force. What is the sum of all these time zones and the effect it has on conservation of motion?

Disclaimer:

The stated "known facts" in this essay are common scientific teachings derived from Albert Einstein's Theories as well as many other famous Physicists. The other ideas put forth here are from my own "thought experiments" and 25 years of fascination with this subject area. I was even "mildly teased" many years ago in Army Advanced Individual Training (AIT), Ft. Gordon, Georgia by some Instructors and Classmates about my theory that gravity is not a force. If anyone has published any of these ideas prior, it is unknown to me and no Plagiarism was intended. I have never found any ideas similar in 25 years about time, have you?

If anyone would like to publish these ideas, then make sure I get notified first and get credit and of course, lots of money. Everyone likes lots of money for their ideas, right?

Do not plagiarize! This is openly published intellectual material! Forensic Writing Analysis works on text too! Patterns are easily determined by style, punctuation, verb and adjective usage, speeling mistakes and Grammar use/misuse!)

Roy Randal Porter Jr. royrporter@live.com, rporterj@nyit.edu
PO BOX 1904
Kingston, New York 12402
Self-Copyright 1999 and forever there after ☺
(Can I do that?)

ROY PORTER JR.
rporterj@nyit.edu

NYIT Summer Session I, July 01, 2010
WRIT 316, Technical Communications Course

Proposal:

Further Studies on Gravitational Time Dilation:

Expanded Tests and Future Applications of Einstein's Theories

Summary

Science is continuously testing, re-proving and examining the effect of Gravitational Time Dilation. This proposal may add insights and possible technological applications to both Einstein's General and Special Theories of Relativity as described in "The Principle of Relativity" (1952) and "The Meaning of Relativity" (1956). There has always been an apparent similarity between the two theories but *why* they are similar still seems to need to be researched further. The proposed research will accomplish many things to possibly include furthering knowledge and real world applications. The specific goals are:

1. Determine to what scale Gravity will act equally in a vacuum
2. Determine what effect distance has considering both Gravity and Time Dilation
3. Develop Technology that benefits from Time Dilation

Introduction

1. Einstein's Theories of General and Special Relativity have been tested and proved repeatedly. These tests always showed that the two different theories always had one common factor, Time Dilation, but caused by two seemingly opposite forces; Acceleration, without Gravity and Acceleration caused by Gravity. In both cases, Time Dilation is present at varying rates as well as the force of gravity due to mass or due to acceleration.
2. Physics has used the velocity equation throughout science with a standard second that is unchanged. Time Dilation changes the value of a second, which is not calculated in? What, if any relevance does this extremely small change in the value of a second have in larger scale experiments?
3. To date, experiments on these factors have been done on what would be compared to a submicron-scale even though we are dealing with "Cosmic" size forces. Modern technological advances and research can now test the same theories on a more "Cosmic" scale in a "Cosmic" environment.

Project Description

1. Determine to what scale Gravity will act equally in a vacuum

Using NASA's ideas for faster, cheaper and better and The Planetary Society's Micro Rovers (Micro Probes in this application), proposal one and two will be combined using the same space probes.

To first test the scale of gravitational effects in a vacuum on a larger scale, a simple probe equipped with a solar sail will be deployed within safe limits to the sun.

All probes may consist of only needed basic components which may include tethers, tether tension detectors, telescoping radial with tension detectors, atomic clocks, oscillators, transmitter / receivers, lasers/receivers and computer modules. Off the shelf computer and electronic components can be modified for cost reduction and the solar sail would act as partial radiation shielding, reducing the costs further.

The "mother" probe would maneuver so that it is at "station keeping" using the solar sail and the sun's gravity. Here, various "experiments" may be released that will immediately react to the suns gravity once free of the probes sun sail equalization.

One experiment would be to release two very different mass objects, each equipped with an atomic clock, computer, laser / detector and telemetry. The laser and telemetry would monitor the fall to the sun of both objects and any changes in Time Dilation using the atomic clocks as a reference. Laser Pulses between the objects and probe would relay digital information, test for red shift and back up radio telemetry as well other added applications.

The next two experiments would involve the use of 3 sets of identical objects, each equipped with small scale telemetry etc. The first set is released individually, one slightly after the other, free space between. The second pair are "soft" tethered together and the last two are "hard" tethered together (telescopic rigid tether). This would give 3 different effects, once again monitored by an atomic clock, laser rangefinder and telemetry both digitally embedded with the atomic clock time and data.

2. Determine what effect distance has considering both Gravity and Time Dilation

This phase would be based on the same technique as the LIGO detector (www.ligo.caltech.edu). Additional probes would be deployed and tethered to a "mother" probe similar to the probe used in phase 1. The tethers, with tension detectors on all, would be as long as possible (Kilometers?) and exactly measured to the millimeter. Once again using NASA tested technology. Solar Sails would also be used on each probe to unfurl and maintain the tether length. Each probe would be equipped with the same atomic clocks, lasers, basic computer and telemetry to communicate with all other probes.

Once all probes are in position and checks done, the solar sails would be adjusted to propel the entire "array" away from the sun. Continuous measurements of distance, tether tension, time and velocity would be relayed to scientists for analysis. They would specifically be looking for differences in length (Special Theory), differences in time rate (both theories) as well as other tests that may be proposed by NASA, Colleges, Universities or organizations. This phase would only begin after the previous experiments were released.

3. Develop Technology that benefits from Time Dilation

This proposal would create a platform that could be reused and easily reconfigured electronically for further experiments and design tests. The initial research and cost to construct would be justified by the future research, experimentation and applications possible. The platform would consist of multiple modern technologies combined to create a test of Einstein's Theory of Special Relativity and possible applications.

Using present materials already developed, a hardened disc that could withstand extreme rotational forces would be built. The disc's circumference should be as large as engineering allows. Multiple size linked discs or arcs may need to be engineered to reduce overall weight and rotational forces. To further reduce overall weight other designs including spokes between circular levels may be preferable. Multiple levels would increase experimentation possibilities. This disc would be engineered to incorporate "Mag Lev" Technologies for both a frictionless magnetic bearing as well as acceleration. The containing bearing/acceleration module would also enclose the suspended disc in a vacuum.

Along the disc(s) at intervals from center to the outer edge would be raised circumference ridges or platforms that would be used to "hold" micro scale electronics. The centrifugal force would only press the components outward into the platform, securing them even further than gluing or other form of securing.

These "electronic packages" would use already developed advanced small scale and Surface Mounted Electronics and be easily redesigned and replaced as units on the disc. Each "package" would contain internal oscillators, small digital processors, memory capability, and wireless transducers similar to RF tags. The units could generate their own power using the rotation and external magnets.

Radial "external" connections can also be engineered from the center of the disc to the various "Package sites" for further connectivity and power. Rotating the placement of the "packages" within each platform would also see if electron flow or functionality is affected at different angles to spin.

This whole disc would be rotated to try to create Time Dilation as the multiple levels of the disc approach a small percentage of "relativistic" speeds. Each "package" would have its own internal clock and also receive external "clock pulses" from transducers based entirely on the rotation of the disc. The digital processors would then record both and a comparison could be made between the two clock sources.

Static identical packages would placed at various distances and locations off the disk not only as a reference but to see if any induced

"Time Dilation" radiates away from the disc. If external Dilation is detected, then according to the equation of velocity and gravity, this external change in time may also create artificial gravity or anti-gravity but at a minute scale. The length of a second would change, therefore changing the ratio in the equation of velocity by changing the value of the denominator.

Methodology and Timeline

Due to the mutual interest in these topics at Universities, NASA, The Planetary Society and other Scientific Organizations and Groups, proposing an "open" involvement would provide extensive support, development and funding. Any contributions of ideas, research and funding would be mentioned in any scientific reports or journals. Additionally any discoveries and patents would be shared according to contributions of the same.

Other organizations such as BOINC from Berkeley can be asked to participate in networked computing of all data received. Other possible contributors may be; Government, Military, Private and Aero-space industries in the applications of the proposed research.

Qualifications

I am a third year Computer and Electronics Engineering Technology Major at the New York Institute of Technology. I have over 14 years experience as a Lead Technician in Satellite, Microwave and Telecommunications systems with the US Army. I have also electronically published a paper "The need for a Universal Time Constant" which has been sent to various scientific organizations including NASA, The Planetary Society and Discussion Boards.

References

Arakaki, Genji "New Millennium DS-2 Electronic Packaging Smaller, Faster With "Managed" Risk" 1998

California Institute of Technology, Jet Propulsion Laboratory http://trs-new.jpl.nasa.gov/dspace/bitstream/2014/22782/1/97-1313.pdf

Betts, Bruce "Microrovers, A New Project: Microrovers for Assisting Humans" 1993-2010 The Planetary Society. http://www.planetary.org/programs/projects/microrovers/

California Institute of Technology. Massachusetts Institute of Technology. "Laser Interferometer Gravitational-Wave Observatory.2001 http://www.ligo.caltech.edu/LIGO_web/about/index.html

Einstein, Albert "The Special and General Theory" 1961 Estate of Albert Einstein

NASA" The Space Tether Experiment" 2001 http://www-istp.gsfc.nasa.gov/Education/wtether.html

The Planetary Society. "What is a solar sail?" 1993-2010 http://www.planetary.org/programs/projects/solar_sailing/whatis.html

University of California. "Berkeley Open Infrastructure for Network Computing (BOINC)" 2003 http://boinc.berkeley.edu/

ROY PORTER JR.
rporterj@nyit.edu

NYIT Summer Session I, July 05, 2010
WRIT 316, Technical Communications Course

Proposal:

Further Studies on Gravitational Time Dilation:

Expanded Tests and Future Applications of Einstein's Theories

Mechanical and Engineering Description

CONTENTS

The following pages are proposed engineering designs as a supplement to the Research Proposal:

Further Studies on Gravitational Time Dilation: Expanded Tests and Future Applications of Einstein's Theories.

Section 1

Introduction

This section describes possible Main and Supplemental Space Probe designs to include component list and structural requirements. Modifications and other designs will be accepted and considered. Final design and payload will be determined by committee and available funding for the project.

Main Probe

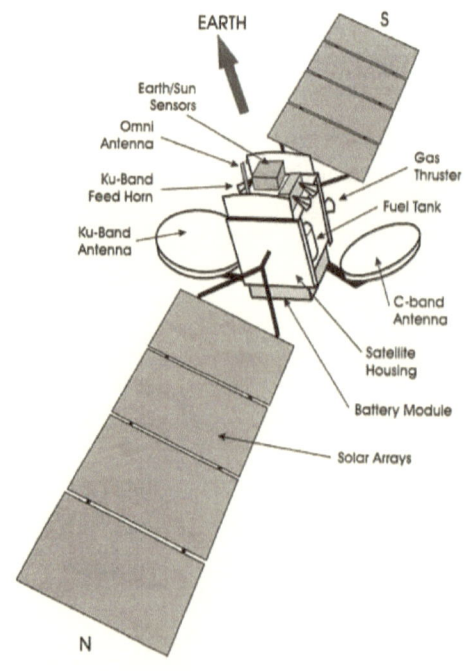

1.1 Main Probe

The main probe would follow basic communication satellite designs with the addition of a forward solar sail and any needed heat and radiation shielding. Any Solar Panels would be engineered to either extend beyond the diameter of the solar sail or more feasible, be positioned in front.

The primary purpose of this probe is communications. It will have multiple band transponders for telemetry and data exchange with all supplemental probes as well as multiple laser transceivers. It would also have long range communications capability for data exchange with Earth.

This probe would also include all necessary computer, guidance and integration systems. The main computer would be larger than that of the supplemental probes with greater data storage to await transmission to earth. It would also contain major component backup systems, reprogrammable hardware and external ports for future applications.

Since this probe will be in station keeping near the sun, its design should be such that many different experimental and sensor platforms may be added or removed by future probes with a simple docking maneuver. This would justify the design and launch costs foreseeing future uses. Propulsion would be accomplished simply by adjusting the size and orientation of the solar sail along with the gravitational pull of the sun.

1.2 Supplemental Probes

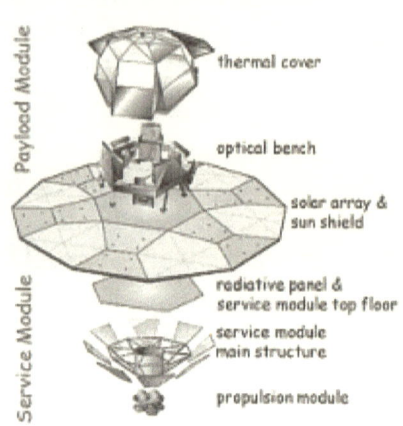

Each Supplemental Probe would require a forward Thermal Cover,

Optical Bench is for sensors and any special experimental devices.

Below that would be the solar array and sunshield. The Solar array would also supply solar electricity as well as act a sun shield.

The radiative panel radiates probe and component heat to space.

The Service Module houses all sensitive electronics and devices.

The propulsion module would be for any needed course corrections and to maintain orientation of the probe.

Each Supplemental probe will be equipped with:

Radio Telemetry-Short range for communication with main probe.

Basic Guidance System-Course correction and to maintain directional orientation of probe.
Laser system-Vectorable transmitter and receiver for additional communication and tests.
Main Computer-Minimal off the shelf technology
Clock- Atomic and Oscillators interfaced with main computer module for experiments.
Interface Computer-Integrates all systems operations and transmitted and received data.

Additional structures and components will be added as needed according to the function of each individual Supplemental Probe. Tether mounts, reels with release system and tension sensors are also added if needed.

A scaled down version of this probe may also be used in the initial gravity test and attached to the different size masses to be released.

This probe in itself may be sufficient alone for the 3 additional release experiments.

A small solar sail will be attached to the probes for LIGOS detector based experiments.

Section 2

Introduction

This section describes some mechanical and engineering aspects of the proposed experiment on Einstein's Theory of Special Relativity. The goal is to engineer a platform using modern technologies to further test and explore possible applications of this theory. Past experiments were limited by computer, electrical and mechanical engineering knowledge of that time. Advances in all areas since the last known test have been extensive and when combined may result in a better understanding of the Theory and possible applications.

2.1 Platform Description

The main platform has a few possible design configurations. The first design configuration would consist of a disc with as large a diameter as engineering allows. Diameter, thickness and surface features would need to be balance to create a precise weight and distribution of weight to achieve smooth rotations at extremely high speed.

Engineers would be required to research the materials available that could withstand extreme forces then determine the largest size possible considering the addition of small electronic packages placed in recesses engineered in the disc. These recesses would be uniform in size and positioned so that centrifugal force would assist in holding any packages added.

Another additional requirement would be the addition of magnetic areas or magnetic coils to create both the frictionless magnetic bearing and for acceleration through magnetic means. This whole disc would be magnetically suspended in a sealed vacuum enclosure to prevent any problems with air movement. This enclosure would also be engineered to contain the bearing system, magnetic acceleration system and various electronic packages.

The second design could instead be a large ring engineered in a similar fashion, but enclosed in a ring shaped contained, also engineered similar to the previous platform.

Any platform would also be mounted so it could be rotated 360 degrees on any axis for tests at different angles to Earth's own gravity. Also, the entire platform will be monitored at its anchor points for any minute change in overall weight.

2.2 Electronic Package Description

The electronic package design can be as extensive or basic as needed. Each package would be designed to be "fitted" into the recesses on the rotational platform and on various locations on the containing platform. Each package would fit into any engineered slot and can be replaced with an identical size upgrade or new design.

On the rotational platform, weight and placement would need to be precisely controlled to prevent disc unbalance. Each "slot" should be large enough to fit a small electronics package that is precisely balanced and weighed. Epoxy etc. can be used to add weight or balance. Each package would also be designed to be able to be rotated in any direction for flexibility in data received at different angles to rotation or differences due to electron flow, gravity or other interferences.

The rotational platform packages would contain as a minimum:

Precision Clock Oscillator
Programmable Microcomputer Chip with Memory
Small Scale Radio Transceiver for data
Receiver for second Clock based on rotation. (Magnetic induction from fixed point?)
Comparator circuits for Clock

The Containing Platform can be fitted with any combination of electronics. The majority would serve to compare internal Clock Oscillations with Clock data from the known rotational speed. This would serve as a control as well and detects any changes in time. Distance of these packages from the rotational platform would also be varied for additional data.

Addition July 15, 2010

The entire platform could also be used to study electromagnetic forces and possible applications of magnetism. We all know that magnetism alone is not a force that can be used as propulsion as is. What if by applying more magnetism to the bottom of a high speed rotating platform results in that force being moved to another area by the spin, creating a force that can be harnessed.

Both the containing platform and rotating platform should also be equipped with controllable electromagnet areas for further tests in magnetism. The sensor packages could also be modified for such tests. Would varying the magnetism up, down, left or right cause any force in the opposite direction? This would be true field propulsion. Add to that the possibility of time dilation changing the denominator in the velocity equation (The value of one second changes). If one side of any mass has a different velocity then what would the results be (movement to equalize conservation of motion?) and could that be electronically controlled?

About the Author

I was Born on May 31, 1964 in Bridgeport Connecticut and at the age of 7 my family returned to his Mother's home town of Historic, Kingston, The First Capital of New York State until burned by the British in 1777 right after our Declaration of Independence. My Father is Roy Randal Porter Sr. whose ancestry originated in England. His Father was an Engineer with General Electric in Connecticut. My Mother's Family emigrated through Ellis Island and settle in the Polish Sector in Kingston New York in the 1800's.

I can personally blame my parents for this book since before I was the age of 7 they began having "Star Trek" TOS family and friends night to watch the newest episodes. Due to their lack of censoring me from this influence, I became addicted to Star Trek and every spin off as well as anything Science Fiction. After they illegalized Star Trek, my addiction of choice became Stargate SG-1, Stargate Atlantis and Stargate Universe with occasional indulgence of Eureka, V, Battlestar Galactica, Caprica and other mind altering TV shows.

In High school I started reading books by Einstein as well other Famous Physicists and developed another mind altering addiction called "Thought Experiments". I began a second addiction of reading Science Fiction and Hard Physics Book even when I couldn't understand the math. This led to hours of "thought experiments" and an over active Science Fiction imagination. Then somewhere in the 1990's I began making up my own ideas and running and rerunning them as thought experiments, which led to this book.

I graduated Kingston High School in 1982. After years of working in the electronics industry I decided to join the US Army in 1992 as a Satellite and Microwave Systems Technician. I served in 2 Hazardous Duties zones while stationed in Panama, Central America. After Panama, I served another two years at the US Army's Electronics Proving Grounds in Ft. Huachuca Arizona.

While serving two tours or 7 years in Darmstadt, Germany, the home of the REAL Frankenstein Castle, I was deployed with 22nd Signal Brigade and became an Operation Iraqi Freedom War Veteran stationed at Mortaritaville, Balad, Iraq. I returned to Iraq once more but this time to Camp Victory, Baghdad and then received an Honorable Medical Discharge at the Rank of Staff Sergeant in 2007.

I then worked as a Mainframe Test Technician at IBM, Poughkeepsie while returning to school. Soon after the Post 9-11 GI Bill was enacted, I returned to school full time and I am still a Student today.

I have attended numerous colleges before and while serving in the Army such as Cochise College, Florida State, Central Texas and Ulster County College. I am presently majoring in Computer and Electronics Engineering Technology at the Manhattan Campus of the New York Institute of Technology.

www.ingramcontent.com/pod-product-compliance
Lightning Source LLC
Chambersburg PA
CBHW020348290526
45785CB00005B/2194